ROUTE

MAINE

BY VIRGINIA M. WRIGHT

Down East Books

MAINE

Cover design by Miroslaw Jurek

Interior design by Lynda Chilton

Printed in the United States

Down East Books

www.nbnbooks.com

Distributed to the trade by National Book Network

CONTENTS

MANY NAMES, ONE HISTORY

by EDGAR ALLEN BEEM

U.S. Route 1 in Maine is 527 miles of pavement that snakes its way from Kittery to Fort Kent, the northern terminus of the historic road that begins (or ends) 2,390 miles south in Key West, Florida. Though Route 1 is old, established, and familiar, it is also a dynamic highway that refuses to lie quietly in its bed. It's always on the move, often at the center of controversy.

The initial construction of what would become Route 1 can probably be traced to 1653 when the Crown Commissioners of Massachusetts ordered that decent roads be built so they could get to the

Postcards tell Route 1's story through landmarks. Some, like Freeport's Indian, have changed little; others, like the Ogunquit Motel, a lot. Prospect's Sail In and Anchor is no more, while Kittery's sardine man is now a lobsterman in Prospect Harbor.

5

Province of Maine to hold court. Towns naturally balked at building what became known as the King's Highway, so it remained wheel ruts and horse paths well into the eighteenth century.

When Maine became a state in 1820, it inherited five turnpikes, sections of which would become parts of Route 1 — the First Cumberland Turnpike in Scarborough, Bath Bridge and Turnpike, Wiscasset and Woolwich Turnpike, and the Camden Turnpike.

The official designation as U.S. Route 1 took place in 1926, right in the middle of Maine's early-twentieth-century road building boom. Between 1914 and 1935, the State of Maine spent $134 million building 1,427.86 miles of state highways as Maine was entering the automotive age. "There is a growing sentiment in many sections of the state," wrote Maine's highway commissioner Paul D. Sargent in 1909, "that the future development of our tourist and summer resort business depends largely upon the development of our system of trunk line highways."

During the 1911 State Senate debate on establishing main trunk line highways, however, not everyone was as bullish on road building as Commissioner Sargent. Senator W.M. Osborn of Pittsfield, for one, found the highway plan "a rash and uncalled for proposition."

"Only a small part of the people of Maine ride in automobiles, less than 3 percent," Osborn argued, "and only a very small portion of those who do use automobiles will be able to travel very much on the trunk line of highways. . . . The people who work for a daily wage and the farmers of the state will not receive practically any benefits from the road."

Governor Percival Baxter, on the other hand, could clearly see the motorized future in 1923 when he spoke at the annual banquet of the Maine Automobile Association

"Just Rolled In." Underwood Motor Camp, Falmouth Foreside, Near Portland, Maine.

"There is a growing sentiment in many sections of the state that the future development of our tourist and summer resort business depends largely upon the development of our system of trunk line highways."

— PAUL D. SARGENT, MAINE HIGHWAY COMMISSIONER, 1909

> "No branch of the state's work is more important or more vital than that of providing suitable highways for our people and our industries."
>
> — GOVERNOR PERCIVAL BAXTER, 1923

THE MANY NAMES FOR ROUTE 1			
West Main St.	Kendall St.	Downeast Hwy.	Park St.
East Main St.	Military St.	High St.	Payne Ave.
Frenchville Rd.	Court St.	Bucksport Rd.	New County Rd.
Main St.	Calais Rd.	Acadia Hwy.	Atlantic Hwy.
State St.	Houlton Rd.	Searsport Ave.	Bath Rd.
Caribou Rd.	South Rd.	Northport Ave.	Leeman Hwy.
Van Buren Rd.	Baring St.	Atlantic Ave.	Mill St.
Presque Isle Rd.	North St.	South Belfast Rd.	Pleasant St.
Houlton Rd.	River Rd.	South High St.	Old Portland Rd.
Benjamin St.	Dublin St.	Elm St.	Portland Rd.
Military Rd.	Cherryfield Stretch	Commercial St.	York St.
U.S. Route 1	Milbridge Rd.	Camden St.	Post Rd.
North Rd.	North Main St.	Union St.	State Rd.
			Island Ave.

The Memorial Bridge carried Route 1 across the Piscataqua River between Badger's Island in Kittery to Portsmouth, New Hampshire, for eighty-eight years. After being closed twice for emergency repairs, the truss lift bridge was permanently closed to traffic in 2011 and dismantled in 2012. Its replacement is expected to open in the summer of 2013.

in Portland. "In 1913," Baxter said, "good roads were few and far between. Today, it is universally agreed that there is no branch of the state's work more important or more vital than that of providing suitable highways for our people and our industries."

The governor was pleased to report that the Kittery-Portsmouth Memorial Bridge, the lift truss span that would carry New England Interstate Route 1, also known as the Atlantic Highway, across the Piscataqua River from New Hampshire, had been completed that very year. Extrapolating from the 46,450 cars that crossed the bridge in its first week of operation, Governor Baxter speculated that 1,388,970 cars a year would cross the new bridge. (Memorial Bridge was closed for reconstruction in 2011, temporarily displacing Route 1 to the Sarah Mildred Long Bridge.)

---

66 Route 1 is a road in constant transition. In a lot of states, Route 1 has lost its interstate transportation reason for existing. Here in Maine, it's kept its identity, but it's different in every geographic part of the state. 99

— JOHN DORITY, FORMER CHIEF ENGINEER OF THE MAINE DEPARTMENT OF TRANSPORTATION, 2012

The route that U.S. Route 1 follows was largely determined in the 1920s by E.W. James, chief of design for the federal Bureau of Public Roads. In a letter dated February 21, 1967, James described how he selected the route the road would take: "I got at once in touch with [Paul Sargent, Maine's highway commissioner] on my proposed numbering scheme," wrote James, "putting pressure on my proposed Route No. 1, which I suggested as the first road along the Atlantic side of the U.S.A., following as far as possible the old, historic Falls Line roads. As soon as I mentioned the Falls Line Route, Sargent said he was with the whole idea, and that the Falls Line Route really began up in Maine, at Fort Kent on the Canadian border."

The Falls Line marked the farthest point a cargo ship could penetrate the coast before running into falls and rapids, necessitating that goods be transferred to wagons. So U.S. Route 1 would initially follow the Falls Line where docks and warehouses and coastal towns had long been established.

Contemporary Route 1 is a contentious roadway, changing names and routes and its very nature from mile to mile and year to year. Route 1 has often been in the news because of spirited debates over whether to widen 1.6 miles of the roadway through the midcoast town of Warren, where a dozen anti-widening protesters were arrested in 2002, and whether to bypass scenic downtown Wiscasset, a well-known Route 1 bottleneck in the summer. The Maine Department of Transportation gave up on the

Wiscasset bypass in 2011 after decades of studies, discussions, meetings, and proposals.

Change has been a constant, whether it's just a name change, such as designating the section from Ellsworth to Danforth as Downeast Coastal Route 1 in 1998, or an actual route change, such as that through Portland in 2007.

Route 1 in Portland used to come in over the Veterans Memorial Bridge and then proceed down Valley Street, turning right onto Park Avenue, left onto Forest Avenue, and right again onto Baxter Boulevard before heading north out of the city. In 2007, however, U.S. Route 1 was relocated from surface streets to run over Interstate 295, an administrative nip and tuck that shortened America's number one highway by a full 1.2 miles.

In parts of Aroostook County, Route 1 is a secondary road. In Camden and Wiscasset, it is Main Street. Between Brunswick and Bath, it is a four-lane highway. And in York County it can be a slow-moving tourist trail.

"Route 1 is a road in constant transition," says John Dority, who retired in 2009 as the Maine Department of Transportation's chief engineer after a career of fifty-four years. "In a lot of states, Route 1 has lost its interstate transportation reason for existing. Here in Maine, it's kept its identity, but it's different in every geographic part of the state."

SHIFTING SIGNAGE

When automobiles first chugged along Maine roads, the routes were marked not by names or numbers but by color bands, much as hiking trials are today. The Atlantic Highway from New York to Calais via Portsmouth, Portland, Rockland, and Bangor was marked with blue bands. An "R" or an "L" was painted on the banded poles at intersections to mark the way.

In 1913, Maine adopted a lettering system to designate major roads, so that the road from Kittery to Portland became State Highway A, from Portland to Bath State Highway C, from Woolwich to Belfast State Highway D, and Belfast to Bangor State Highway L. U.S. Route 1, following the Falls Line, originally went up the Penobscot River to Bangor.

In 1925, Maine transitioned to a number system with the first state road markers painted yellow with black numerals. The U.S. Route 1 shields, inspired by the Great Seal of the United States, have always been black and white, but when they first went up in 1927 they also contained the name of the state.

In 1947, Route 1 was designated a Blue Star Memorial Highway "to perpetuate the memory of the men and women of Maine who served in the armed forces of the United States in World War II."

THE ROAD LESS TAKEN

by EDGAR ALLEN BEEM

The refusal of U.S. Route 1 to lay still is nowhere more apparent than in the diversions it makes as Route 1A, Business 1, and Bypass 1, alternative routes, known as bannered routes, that sometimes follow their original course and sometimes just make side trips around or downtown. What they all have in common, though, is that they eventually return to the main road.

There are 1A tangents along the summer shore from York to Cape Neddick, along the Portland waterfront, a Business 1 swing through Damariscotta and Newcastle, a Route 1 bypass around downtown Rockland, a 1A stretch from Milbridge to Harrington while Route 1 wanders up to Cherryfield, and another from Jonesboro to Machias through Whitneyville. And then there are the two major 1A detours — the fifty-mile sidetrack from Mars Hill to Van Buren as Route 1 in Aroostook County veers west through Presque Isle and Caribou, and the fifty-mile loop from Stockton Springs up to Bangor and back down to the coast at Ellsworth.

Aroostook County's 1A route evolved over the course of several decades as Route 1 and Route 1A between Mars Hill and Presque Isle essentially switched places, 1A running through Easton after Route 1 abandoned the Easton route. For years, Route 1A ended in Caribou. Its current routing was established in 1989 when Representative Hilda Martin (D-Van Buren) succeeded in persuading the federal government to designate Maine's Route 165 from Fort Fairfield through Limestone and Caswell to Van Buren as U.S. Route 1A in order to improve the roadway along the border.

The Bangor 1A loop is a function of the fact that U.S. Route 1 originally followed the Penobscot River from Stockton Springs up to Bangor and then back down the other side of the river to Ellsworth. In 1930-31, however, the scenic Waldo-Hancock Bridge across the Penobscot was built, connecting the towns of Prospect and Bucksport over Verona Island. The new bridge, dedicated in 1932, reduced the drive from Belfast to Ellsworth by thirty-seven miles and an estimated forty-five minutes. (The Penobscot Narrows Bridge replaced the Waldo-Hancock Bridge in 2006.)

In the 1960s, John Dority, the Maine Department of Transportation's chief engineer, oversaw the designation of Business 1 through Damariscotta and Newcastle and the design of the new Route 1 bypassing those downtowns. Back then the relocation faced little public opposition.

"The only person absolutely against it was Bob Reny of Renys Department Store," recalls Dority. "But Damariscotta was a worse chokepoint than Wiscasset ever was. There was one public hearing, and we were building within a year. Of course, we couldn't build there today."

Maine people are much more sensitive to the state's natural and cultural heritage than they were fifty years ago. Just as the presence of an eagle's nest helped defeat the long-sought Wiscasset bypass, the fact that the new route of Route 1 around downtown Damariscotta and Newcastle went right through an ancient Indian shell midden dating to the Red Paint People would be a road killer today — chokepoint, bottleneck, traffic jam, or no.

BY THE NUMBERS: MAINE ROUTE 1

(These figures are open to dispute, but you'll have to drive the 527 miles and do your own counting.)

21 McDonald's restaurants

96 churches

1 wind farm

4 degree-granting colleges or universities

28 cemeteries

4 border crossings within view of Route 1

25 self-storage facilities

3 auction houses

2 islands

2 hubcap yards

8 moose crossing signs

4 signs featuring fishermen in yellow oilskins

41 Irving stations

58 auto sales dealers

14 Family Dollar stores

5 Grange halls

1 purple house

2 state parks

3 stop signs

1

THE MAINE ROAD

Fort Kent

Frenchville

Madawaska

Grand Isle

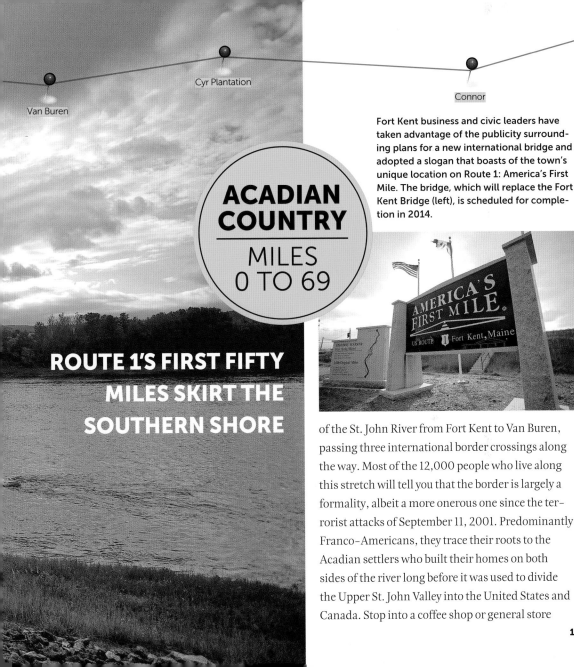

Van Buren

Cyr Plantation

Connor

Fort Kent business and civic leaders have taken advantage of the publicity surrounding plans for a new international bridge and adopted a slogan that boasts of the town's unique location on Route 1: America's First Mile. The bridge, which will replace the Fort Kent Bridge (left), is scheduled for completion in 2014.

ACADIAN COUNTRY
MILES 0 TO 69

ROUTE 1'S FIRST FIFTY MILES SKIRT THE SOUTHERN SHORE

of the St. John River from Fort Kent to Van Buren, passing three international border crossings along the way. Most of the 12,000 people who live along this stretch will tell you that the border is largely a formality, albeit a more onerous one since the terrorist attacks of September 11, 2001. Predominantly Franco-Americans, they trace their roots to the Acadian settlers who built their homes on both sides of the river long before it was used to divide the Upper St. John Valley into the United States and Canada. Stop into a coffee shop or general store

15

anywhere along this route and you are likely to overhear conversations shifting seamlessly from French to English to French. Faith, too, defines the culture here. Crosses and Virgin Mary statues are liberally displayed on lawns and doorsteps, and ornate Catholic churches dominate the otherwise humble villages. In Van Buren, Route 1 leaves the St. John River, turning sharply south to ride the open, rolling hills for which Aroostook County is best known.

"Our record is thirty-six below." — John Freeman, who with his wife, Mary, serves up spare ribs, pulled pork, and barbecued chicken from the Rib Truck at locations along Route 1 from Fort Kent to Presque Isle all year round and in all kinds of weather.

Greg Sirois (above) works at the Sled Shed in Van Buren. A man strolls down Main Street in Madawaska (right). Bilingual road signs (below) appear often in the St. John Valley.

In winter, snowmobiles are as common as cars in the parking lot of the Northern Door Inn in Fort Kent.

U.S. Highway 1 has no first mile, and it has no last mile. Not officially, anyway. It only has a pair of neutrally designated termini, Fort Kent, Maine, and Key West, Florida, with 2,390 miles of pavement between them. That's the protocol of the American Association of State Highway and Transportation Officials (AASHTO), which coordinates the country's vast network of U.S. Numbered Highways for the state and local governments that maintain them.

The folks in Fort Kent aren't much for protocol, at least not in this instance. As far as they're concerned, Route 1 begins here. The town's slogan, adopted in November 2010, is "America's First Mile," and it is prominently displayed on a large granite sign near the international bridge that connects the town of 4,000 with Clair, New Brunswick. (The marker replaced a wooden welcome sign that had similarly marked "the beginning of Route 1" for years.)

History supports Fort Kent's claim. "U.S. Route 1 became the first designated U.S. Route in 1927,"
says Carl Pelletier, manager of the Northern Door Inn (also known as La Porte du Nord), which shares with Rock's Family Diner the distinction of being the first business on Route 1 South. "It started here in Fort Kent and went to Miami. It wasn't until 1938 that they completed the Overseas Highway, which connected lower Florida to Key West and became an extension of Route 1."

While Pelletier's personal opinion is plain, he is a diplomat. He has erected a "Historic Marker" in the Northern Door Inn's parking lot. On one side, the sign bears the familiar Route 1 logo with the declaration: "MILE 0, Fort Kent, ME." The other side appears identical except it asserts "MILE 2390, Fort Kent, ME."

"Not being one who wants to argue, I let you have it whichever way suits your purpose," Pelletier says. "If I had a dollar for every time people took photos of themselves in front of that sign, I'd be long since retired."

St. Louis Cemetery (left) in Fort Kent overlooks the St. John River. The Fort Kent Blockhouse (above left) was built during the Aroostook War (1838-1839), a border dispute with the British in Canada. Rock's Family Diner in Fort Kent commemorates Mile 0 with T-shirts (below).

 PETER PINETTE
OWNER OF ROCK'S FAMILY DINER, 378 WEST MAIN STREET, FORT KENT

"Oui, je parle francais. My mother is from Edmundston (New Brunswick) and both of my parents were bilingual. Years ago Fort Kent was primarily French, and over the years it's transitioned to maybe half and half. Across the river is Clair, a French-speaking part of New Brunswick.

"Rock's has been here since 1945. Rock Ouellette started it as a small hot dog stand when he was in high school, and he expanded it after he got out of the service. I'm the third owner. I bought it in 2001.

"A lot of people come to Rock's because they've known it for so many years. We have heard stories of people from Connecticut who drive all night just to have a hot dog and a mixed fries [poutine], then they turn around and go back home. It's not so much the brand of hot dogs that they like as the way we cook them. We slow-cook them and let them age on the grill a bit.

"Saturday afternoons in winter are busy. The snowmobilers and the mushers who come for the Can-Am Crown International Sled Dog Races in February gather here. And now that the America's First Mile monument is across the street, we have people who stop in because they are driving the Route 1 corridor."

LEO CORMIER
LEO'S BARBER SHOP, MAIN STREET, VAN BUREN

"My mother and father were Canadian, but I was born and raised here. You find that a lot around the Valley. When my daughter was going to school, they told her, 'Speak only English.' The nuns were forcing that. Why would you want to let your culture go? We always spoke French to her at home. She turned around and went to school in Boston and got a job at Mass General Hospital. When French-speaking patients came in, they'd call her over to translate. It's an advantage to know two languages.

"Van Buren used to be a booming little town. When I started this business in the seventies, we'd work until nine or ten at night on Fridays and Saturday. Now I'm done Saturdays at noon. The town really changed when they closed Loring Air Force Base. That was a shocker. A lot of Canadians used to come here, too, but then they started their own shopping malls. "If you don't like a busy atmosphere, this is the place you want to be. But if you're young and you want to earn a living, it's hard. I see it so often where some of my old classmates come up from Connecticut and say, 'We'd love to move back now, but we had to leave to earn a living and now we don't want to leave the grandkids.' But they still come to visit. They love to visit."

A sinful snack of French fries, mozzarella cheese, and brown gravy is a staple on restaurant menus in the Upper Saint John Valley, although its name varies from town to town. "Here, it's poutine," says April Hodgson, the owner of Frenchie's Variety in Grand Isle. "In Fort Kent, they call it 'a mix,' and in some towns they call it a 'godfather'." Across the St. John River in Canada, where the dish was invented, fresh cheese curds, not mozzarella, are used.

License plates on the wall of the Sled Shed in Van Buren.

Madawaska, the country's northeastern-most town, is a mandatory checkpoint on the Southern California Motorcycle Association's USA Four Corners Tour, which challenges riders to visit the four corners of the Lower 48 in twenty-one days. It is the only town with a park dedicated to the circuit — **Madawaska Four Corners Park** on Main Street (Route 1) — and the only town that offers groups of riders a ceremonial police escort.

Many St. John Valley residents have family and friends in Connecticut. In the sixties, seventies, and eighties, as mechanization reduced agricultural jobs in Aroostook County, a large number of St. John Valley residents moved to Hartford, Bristol, New Haven, and other Connecticut cities in search of jobs in that state's rapidly growing industries.

Cyr Plantation is comprised of thirteen thousand acres of forest (much of it woodlots), nine thousand acres of farmland, and 107 people. It is the first of three plantations that travelers encounter as they travel south on Route 1, and one of thirty-four such civil divisions in Maine.

So what is a plantation? "It goes back to when Maine was part of the Massachusetts Bay Colony, and it was intended to be a temporary government to guide a community from an unincorporated township to an incorporated town," says Eric Conrad, spokesman for the Maine Municipal Association, "but they've continued in Maine and uniquely so."

Unlike an unorganized territory (such as Cyr's neighbor, Connor), which is governed by the county, a plantation has a local governing body — the annual meeting where community members approve a budget, set policy, and elect a board of assessors to see to day-to-day operations. Unlike a town, however, a plantation does not have complete home rule; its zoning and planning authority is the Maine Land Use Regulation Commission (LURC).

Cyr voters gather for their annual meeting in the 1934 Governor Brann Schoolhouse, a well-preserved one-room red schoolhouse on Route 1, about one mile from the Van Buren town line. "During the meeting the assessors are elected to carry out the daily operations of the government," says first assessor Danny Devoe. "Taxes are raised and appropriated and voters are registered. There is very little demand for public services, like fire protection, law enforcement, and public roads, which we have under contract with the town of Van Buren."

Cyr's population peaked at five hundred people in the mid-1800s when the Homestead Act of the State of Maine encouraged expanded settlement. In the years since, the population has slowly declined. "We haven't grown, but that might change," Devoe says. "They're building a big new port of entry in Van Buren and that will probably create one hundred to one hundred and fifty jobs. Some of those people have expressed interest in living in the rural areas outside Van Buren."

A retired potato farmer, Devoe says his family roots in Cyr date back to the 1850s. "The other two assessors' families have been here since the 1840s and 1850s," he adds. "The same is true for the clerks and the treasurer. The same families have been governing this community for the last one hundred and fifty years. Not much has changed in Cyr Plantation."

>>>DON'T MISS

SIGHTS

Fort Kent State Historic Site, off Route 1, Fort Kent. The main attraction is a wooden blockhouse built during the so-called Aroostook War, a dispute between the United States and Great Britain over the Maine-Canadian border.

Musée Culturel du Mont-Carmel, Main Street, Grand Isle. 207-895-3339. museeculturel.org. Housed in a former Catholic church, this museum displays Acadian and Quebecois artifacts, such as textiles, furniture, and religious items. The church was built in 1909 by Edmundston contractor Léonide Gagné from plans by architect Theo Daust. The twin Baroque-style belfries are adorned with two archangels blowing trumpets, the works of Quebec sculptor Louis Jobin.

Acadian Village, Route 1, Van Buren. 207-868-5042. connectmaine.com/acadianvillage. Overlooking the St. John River, this cluster of seventeen historic buildings includes a 1790s log home, a railroad station; a general store, a schoolhouse, a blacksmith shop, a gristmill, and a replica of an eighteenth-century log church.

EVENTS AND ENTERTAINMENT

Can-Am Crown International Dog Sled Races, Fort Kent. 207-543-7515. can-am.sjv.net. This March event is a qualifying race for mushers wishing to compete in Alaska's famous Iditarod race.

Ploye Festival and Muskie Derby, Fort Kent. 207-834-5354. fortkentchamber.com. Held every August, the Ploye Festival celebrates a buckwheat pancake that is a staple in Acadian homes. A highlight is the making of the World's Largest Ploye, which is twelve feet in diameter. Held the same weekend, the Muskie Derby challenges anglers to catch the giant muskellunge.

10th Mountain Center Biathlon Events. 207-834-6203. mainewsc.org; 10thmtskiclub.org. The 10th Mountain Center is one of four skiing facilities owned by the Maine Winter Sports Center. It hosts numerous biathlon competitions, including the World Cup Biathlon in 2004 and 2011, and the Paralympic Nordic World Championships in 2005.

Acadian Festival, downtown Madawaska. acadianfestival.com. Held every August, the festival includes a re-enactment of the first Acadian landing in northern Maine, a traditional Acadian supper, a "Soiree du bon Vieux Temps" (Night of Good Old Times), and a parade.

FOOD

Dolly's Restaurant, 17 Main Street, Frenchville. 207-728-7050. Traditional Acadian fare like chicken stew, creton (a spiced pork pâté), and ployes, buckwheat pancakes that are served with maple syrup for breakfast or as flatbread for mopping up gravy at lunch or dinner.

The Rib Truck, alternating between Route 1 locations in Fort Kent, Madawaska, Presque Isle, and Bennett Drive in Caribou. ribtruck.com. Flavorful and juicy ribs, pulled pork, chicken, and baked beans to go.

SHOP

Bouchard Family Farm, 3 Strip Road (at Route 161), Fort Kent. 207-834-3237. ployes.com. For generations, the Bouchards have milled the buckwheat flour that is used in ployes, a crepe-like staple of Acadian cuisine. In addition to buckwheat flour, their products include ploye mixes and the *French Acadian Cookbook.*

All street names refer to Route 1 unless otherwise noted.

Amity · Orient · Weston · Danforth · Brookfield · Topsfield · Waite · Indian Township · Princeton · Baileyville · Baring Plantation

White settlers began planting potatoes on Aroostook County's rolling hills in the early 1800s (left). Honor boxes, like this one at Hubcap Heaven in Littleton, are the norm here (below).

POTATO FIELDS AND DEEP WOODS
MILES 70 TO 213

IN A STATE WITH A REPUTATION FOR RUGGED SHORES

and impenetrable forests, the Aroostook County landscape distinguishes itself with its gentleness and tractability. This is Maine's big sky country, a patchwork of potato fields and small villages, with pastoral views that stretch for miles. Presque Isle, the County's largest city (population 9,692), interrupts the rural idyll with its sprawling strip malls and busy downtown, but the urban flirtation is brief as development quickly drops away and farms once again rule the countryside for the next forty miles. Houlton, where Route 1 intersects with Interstate 95 for the first time, is a jarring milestone, marked by truck stops, motels, and fast-food restaurants,

25

and then a dramatic shift in the landscape. Fir trees close in, creating a dense green corridor, broken occasionally by a house and, here and there, an abandoned gas station whose grounds are littered with old farm tools, woodstoves, and, in one case, a fifties-vintage Chevy waiting at the pump for an attendant who never comes. About midway through this section of highway, in Orient, Route 1 rises to offer a panoramic vista of Grand Lake and the forested mountains of New Brunswick on the opposite shore. This is the Million Dollar View Scenic Byway, eight miles of some of eastern Maine's most stunning scenery.

A man walks the Gateway Crossing in Houlton (right) and a couple poses at the Passamaquoddy Reservation (bottom right).

BRIAN BRISSETTE
OWNER WITH WIFE JANE CAULFIELD OF MORNING STAR ART AND FRAMING, 431 MAIN STREET, PRESQUE ISLE

"Presque Isle is the hub of Aroostook County. It has the largest population (9,692), the most businesses, the largest hospital, the University of Maine at Presque Isle, and the Presque Isle Regional Career and Technical Center. Potatoes, logging, and snowmobiling are the big industries here.

"Jane and I have had this business for seven years. It was her idea. She is the economic development director for the Aroostook Band of Micmacs. She had a premonition in the middle of the night that we were going to open a frame shop, and lo and behold, we did. We went to a school in Connecticut to study framing, but we knew we couldn't make it on framing alone, so we added art supplies and Maine-made crafts.

"I thought we were just going to cater to the students from the university, but I was amazed at the number of artist working in northern Maine. We feature work from eighty-five local artists — Native American baskets, candles, soaps, photographs, and woodcarvings.

"Right now I'm framing a caricature of Senator Susan Collins for the Maine Potato Board. It shows the senator as a knight riding a horse, with a lance in one hand and a flag with a big potato on it in the other. It is for her efforts this past fall to allow schools to keep serving potatoes."

The city of **Caribou** owes its name to a single reindeer that was hunted by a settler named **Alexander Cochran** sometime in the 1850s, according to Dennis Harris of the Caribou Historical Society. Cochran shot and wounded the animal, and his dogs then tracked and killed it by a stream that feeds the Aroostook River. The stream became known as Caribou Stream, and it in turn gave its name to the town that grew up around it.

The christening of Caribou was not a straightforward affair, however. It was first incorporated as the town of Lyndon in 1859, says Harris, adding that the origins of that name are not known. Ten years later, the town's name was changed to Caribou, but less than a month later, the name Lyndon was restored. Sometime between 1874 and 1888 the name was again changed to Caribou by the state legislature, and this time the name stuck.

The name **Lyndon** also endures. An area north of town is referred to interchangeably as "North Lyndon" and "North Caribou," and Lyndon Street runs parallel to Caribou Stream just south of downtown.

For nearly fifteen miles south of Presque Isle, 1,748-foot-high Mars Hill dominates the eastern landscape. The rugged little mountain holds the distinction of being the first place in the United States to be illuminated by the rising sun for half the year (March 25 to September 6). Carved into its slopes are Bigrock Mountain Ski Area's twenty-six trails, some of which do double duty as the International Appalachian Trail, which extends from Baxter State Park through New Brunswick and Quebec to Newfoundland's Belle Isle. And along Mars Hill's four-mile ridgeline: the twenty-eight wind turbines that compose Maine's first wind farm, Mars Hill Wind.

From Route 1, which ranges from seven to two miles west of Mars Hill, the turbines appear delicate and ethereal, but they are in fact massive structures, weighing 20,000 pounds apiece and stretching 262 feet into the sky with slowly turning blades that span 115 feet. The wind farm, which generates 130 million kilowatt hours of electricity annually, has not been without controversy. A group of residents who live near the turbines have complained of noise, health disturbances, and a decline in property values.

BY THE NUMBERS: MARS HILL			
1,748	height in feet of the ridgeline	**20,000**	weight in pounds of single turbine
4	length in miles of the ridgeline		
26	trails at Bigrock Mountain	**262**	height in feet of a single turbine
2	ski trails that are part of the International Appalachian Trail	**115**	diameter in feet of a spinning turbine
28	wind turbines that compose Mars Hill Wind	**130 million**	kilowatt hours of electricity generated by Mars Hill Wind

JODI MOORE
HEAD NIGHT AUDITOR, SHIRETOWN INN, HOULTON (INTERVIEWED AT 6:30 A.M.)

"I've been here since eleven last night. I get the graveyard shift. When I started work here nine years ago, I watched all the young kids who worked this shift come and go. Then I started college, and I asked my boss to let me do the 11 to 7. He said, 'I don't know. I've never had a female work the night shift before.' I said. 'This is nothing. I grew up in the bar business for crying out loud — my parents owned a bar.' He agreed to give it a try. Eight-and-a-half years later, here I still am. I love it.

"Most of the time it's quiet. Sometimes the shift can get lonely, but I have plenty to do — I do the nightly audits, I do the payroll, I check the Internet reservations, I get the next day's reservations together. I feel safe because I'm always seeing the police or the border patrol do a drive-through. Sometimes the border patrol comes in and asks if so-and-so has checked in. Usually, the name that is registered doesn't match up, but the vehicle does.

"With I-95 being right here, we get a lot of truck drivers, a lot of regulars. We also get a lot of snowmobilers. Some arrive on their sleds because they're coming from down country and you can pick up the trail right in back of the motel.

"I moved to Houlton in January '03. I had been living in South Carolina and I was in a relationship that was not going well. I opened a map, closed my eyes, and my finger landed on Houlton. I was taken in like I'd been here all my life. You never feel like a stranger here. You walk in a place, and you're greeted with smile and a hello."

In 2000, Virgil Farrar threw a pair of old shoes up into this maple in front of his home in Hodgdon to amuse his grandchildren. Within days, passersby had added several more pairs. There are now hundreds.

"According to the Federal bureau's report, the road is now in good condition throughout its entire length, although nearly 363 of its 2,321 miles are still unsurfaced. Between the Canadian line and Ellsworth, Maine, a distance of 338 miles, there are 72 miles of earth road, the remainder being improved with surfaces of gravel or higher type."

— U.S. Department of Agriculture press release titled "United States Route 1 is a Highway of History," October 9, 1927

1.4 billion: pounds of potatoes harvested in 2011, which ranks Maine twentieth in potato production in the United States (Idaho is first with 12.7 billion pounds)

$540 million: estimated economic impact of the potato industry in Maine

6,150: number of jobs created by Maine's potato industry

54,000: acres of harvested potato fields in 2011, which ranks Maine fifth in number of acres devoted to potato production in the United States

220,000: acres of harvested potato fields in 1946

66: percentage of Maine potatoes used for processing (french fries and chips)

20: percentage of Maine potatoes used to seed new crops

14: percentage of Maine potatoes sold fresh for home and restaurant use

JEREMY FREY
PASSAMAQUODDY BASKET MAKER, BASKET TREE GALLERY, 195 U.S. ROUTE 1, INDIAN TOWNSHIP

"My family has made baskets for generations. My grandfather just stopped because he has emphysema and can't breathe in the dust anymore. He wove utility baskets, big work baskets, whereas I make fancy baskets. My mother taught me when I was twenty-two, so I've been doing it for ten years full time and every year has gotten a little bit better as far as making a living at it goes. I'm more well known now. People seek me out. I've had one person come to this shop from Manhattan. He heard about me from a gallery owner who collects my stuff.

"I harvest my own brown ash. First, you have to find the right tree. It has to be a certain thickness, it can't have knots, it can't have any diseases, and it can't have any spiraling of the grain. In other words, you need a perfect tree, and brown ash rarely grows perfectly. I'd say I cut only one in one hundred of the trees I look at. Next, I pound the trunk from one end to the other, then I turn it a little and pound it again, and so on. I pound every square inch of tree so the

fibers around the growth rings break and I can pull the growth rings off one at a time. I soak them and hand-split them so the pieces are like ribbons that I cut to the width I need.

"I don't know anyone who weaves like I do. I go further in the degree of the precision than any basket maker I've ever seen — intentionally, because there is competition. There are other weavers who use the same materials and the same basic techniques selling at the same shows I do.

"I price my baskets very high. I've been getting so much publicity that if I didn't set them as high as I do, I'd be so backordered. Last year I won the best of show prize at the Heard Museum Indian Fair and Market in Phoenix and the Santa Fe Indian Market. I also won a $50,000 grant from United States Artists. That's what helped us get this gallery going. This will be the most high-end business this reservation has ever seen, and I hope it will inspire people here as much as it will help us."

MAINE'S TOP 5 **POTATO VARIETIES**

Russet Burbank 38%

Frito-Lay* 15.5%

Snowden 5.8%

Shepody 5.2%

Superior 3.8%

No, Frito-Lay is not the name of a potato. Frito-Lay has its own breeding program encompassing several varieties identified only by number.

BILL KOLODNICKI
REFUGE MANAGER, MOOSEHORN NATIONAL WILDLIFE REFUGE, BARING PLANTATION

"Moosehorn was established in 1937 by FDR. It is one of the oldest wildlife refuges in the country. We're about 30,000 acres in two parts — most of the acreage is here in Baring; the other part is in Edmunds, on Cobscook Bay. Our principal mission is habitat for migratory birds. Moosehorn is famous for woodcock research. We also provide public access for recreation. We get about 50,000 visitors a year who come to hike, fish, and hunt. We allow blueberry picking in August, and in December, we let people come in and take Charlie Brown Christmas trees.

"People around town are always asking me about the eagles. We have a pair of bald eagles who have been nesting here on a platform near Route 1 for more than twenty-one years, and they are one of the most publicly accessible pairs. They typically begin nesting in February, and they lay their eggs at the end of March or early April."

>>>> DON'T MISS

SIGHTS

Caribou Historical Center & Museum, 1033 Presque Isle Road, Caribou. 207-498-2556. The museum includes an 1860s replica one-room schoolhouse, a barn with early farm equipment, a portable saw mill, and exhibits of military artifacts from the Civil War, World War I, and World War II.

The Maine Solar System Model, Route 1 between Presque Isle and Topsfield. pages.umpi.edu/~nmms/solar. Extending more than 40 miles from the Sun to a dwarf planet, this is the largest three-dimensional scale model of the Solar System in the world.

The Northern Maine Museum of Sciences Folsom Hall, University of Maine at Presque Isle campus. 207-768-9400. A natural history museum with exhibits on biology, geology, mathematics, physical science, astronomy, chemistry, forestry and agriculture.

Southern Aroostook Agricultural Museum, 1664 Route 1, Littleton. 207-538-9300. Oldplow.org. Vintage farm implements like the A.E. Mooers Milk and Cream cart on sled runners, artifacts of domestic farm life, and a room dedicated to Aroostook County's signature crop, the potato.

Moosehorn National Wildlife Refuge, Route 1, Baring. 207-454-7161. fws.gov/northeast/moosehorn. Birding, especially bald eagle watching, is a big draw at Moosehorn's 17,200-acre Baring Division. More than fifty miles of dirt roads and trails are open for hiking, skiing, and biking. The refuge's 7,200-acre Edmunds Division also is on Route 1, about twenty miles south between Dennysville and Whiting.

RECREATION

Maine Winter Sports Center, 552 Main Street, Caribou. 207-492-1444. mainewsc.org. Dedicated to reviving skiing, the center has built cross-country trails throughout much of Aroostook County. Its alpine and cross-country skiing facilities include the 10th Mountain Center in Fort Kent and the Nordic Heritage Center in Presque Isle. It also is affiliated with Big Rock Ski Resort in Mars Hill and Quoggy Jo in Presque Isle.

Aroostook State Park, just off Route 1, Presque Isle. 207-768-8341. Maine's first state park is located on Echo Lake. At 600 acres, its most prominent feature is 1,110-foot, twin-peaked Quoggy Jo Mountain, which rises above the surrounding farmlands. Activities include hiking, cross-country skiing, and swimming.

FOOD

Café Sorpreso, 415 Main Street, Presque Isle. 207-764-1854. cafesorpeso.com. Fine dining with a menu that changes weekly. Entrées may include anything from filet mignon to dover sole with a piquant onion and bacon crust to pork tenderloin in a Thai orange–curry sauce.

Grammy's Country Inn, 1687 Bangor Road (off Route 1), Linneus. 207-532-7808. Home-style cooking and huge portions, including double-fisted whoopie pies. Lots of deep-fried offerings, from lobster and rainbow trout to broccoli and fiddleheads.

SHOP

Morning Star Art and Framing, 431 Main Street, Presque Isle. 207-764-1810. Displaying the work of Aroostook County artists and craftspeople, including paintings, photography, jewelry, and Passamaquoddy baskets.

Bradbury Barrel Co., 479 Route 1, Bridgewater. 207-429-8141. bradburybarrel.com. Aroostook County potatoes used to be shipped in the cedar barrels manufactured by this company. When shipping methods changed, the company found new markets for its barrels, tubs, and wood display fixtures.

All street names refer to Route 1 unless otherwise noted.

Million Dollar View Scenic Byway

St. John River

Potato fields

ROUTE 1'S **BEST VIEWS**

St. John River
Madawaska to Van Buren

Potato Fields
Van Buren to Caribou

**Million Dollar View
Scenic Byway**
Weston to Danforth

**St. Croix River and
Passamaquoddy Bay**
Calais to Perry

Schoodic Scenic Byway
West Gouldsboro to
Hancock

Camden Hills
Northport to Lincolnville

Clam Cove
Rockport

Wells Beach
Wells

St. Croix

Lincolnville Beach

Wells Beach

Calais

Robbinston

Perry

Pembroke

Dennysville

Edmunds

Whiting

East Machias

Machias

Jonesboro

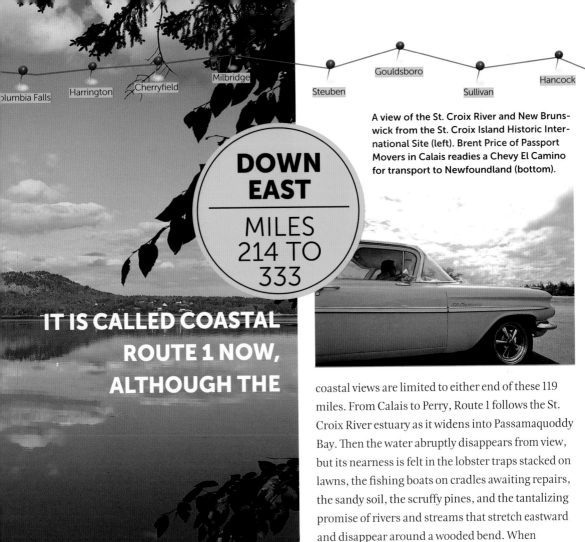

Columbia Falls   Harrington   Cherryfield   Milbridge   Steuben   Gouldsboro   Sullivan   Hancock

DOWN EAST

MILES 214 TO 333

IT IS CALLED COASTAL ROUTE 1 NOW, ALTHOUGH THE

A view of the St. Croix River and New Brunswick from the St. Croix Island Historic International Site (left). Brent Price of Passport Movers in Calais readies a Chevy El Camino for transport to Newfoundland (bottom).

coastal views are limited to either end of these 119 miles. From Calais to Perry, Route 1 follows the St. Croix River estuary as it widens into Passamaquoddy Bay. Then the water abruptly disappears from view, but its nearness is felt in the lobster traps stacked on lawns, the fishing boats on cradles awaiting repairs, the sandy soil, the scruffy pines, and the tantalizing promise of rivers and streams that stretch eastward and disappear around a wooded bend. When the ocean reappears seventy-five miles later in Gouldsboro, it does so spectacularly, affording views of Frenchman Bay and the mountains of Acadia National Park on Mt. Desert Island.

You are halfway between the Equator and the North Pole.

Glimpses of the ocean are limited between Perry and Gouldsboro, but Coastal Route 1 nevertheless lives up to its name with a look, pace, and culture that is distinctly different from the 213 miles that precede it. When the road reaches Ellsworth, it changes yet again.

Sitting on the Calais-Robbinston town line, the St. Croix Island International Historic Site commemorates one of the earliest European settlements in North America. In 1604, an expedition led by French explorer and merchant Pierre Dugua settled on St. Croix Island in Passamaquoddy Bay. Thirty-five of the seventy-nine settlers did not survive the winter. The survivors departed the following summer and founded Port Royal, Nova Scotia.

One of three National Park Service units in Maine (the others are Acadia National Park and the Maine portion of the Appalachian Trail), St. Croix Island International Historic Site is managed under a unique partnership between the United States National Park Service and Parks Canada. A trail with interpretive panels and bronze statues of the settlers offers views of the 6.5-acre island, which sits in the mouth of the St. Croix River.

DONNIE SMITH
WASHINGTON COUNTY SHERIFF

"It would take you two-and-a-half to three hours on a good day to drive the coastal route from the Hancock County line to the Aroostook County line. It's a hike. Washington County is a big county, and there are only three officers covering it. And the coverage is only until midnight. From midnight to eight in the morning, there's nobody covering this county. If something happens, a deputy or a trooper will get called out, of course, but we don't have twenty-four-hour police coverage in this county.

"We call-share with the Maine State Police. Today, for example, there might be two deputies and one trooper on duty, and tonight there might be two troopers and one deputy. We divide the county into three sections or slots. The largest slot goes from Robbinston all the way to the Aroostook County line, but it's different from the others in that it is less populated, there are fewer side roads, and the towns of Calais, Baileyville, and Indian Township have their own police departments. Still, it's a phenomenal task to cover it all.

"It's most daunting when you have a snowstorm — the first and last storms of the season are always the worst. You start hearing traffic accidents, and you know that deputies and troopers are backed up trying to cover them all. Someone could wait two or three hours before you can get to them."

LEWIS SCRIBNER
CO-OWNER WITH WIFE
KAREN OF KAREN'S DINER,
439 MAIN STREET, CALAIS

"I was born in Calais. I left home at age 17 and took a year in Freeport working in a shoe factory. Then I moved to Syracuse and worked in drywall. I came back to Calais in 1972. There never was a day that I was out of the state of Maine that I didn't think of it. So when the opportunity to come back presented itself, I took it. My goal never was to be rich. Home was the attraction that can't be beat. My goal was to support my family, put a roof over their heads, and feed them."

Twelve granite stones on the east side of Route 1 mark the miles from Calais to the Mansion House in Robbinston (near Grace Episcopal Church and Robbinston Visitors Center). The numbered stones were installed in the late nineteenth century by the house's second owner, James Shepherd Pike, Abraham Lincoln's ambassador to the Hague. A racehorse enthusiast, Pike used the markers to clock his horses on his way to his law office. All of the stones are gray granite except for number six, which is red granite and believed to be a replacement for an original that was destroyed during a road construction project.

No one should pass through Machias without sampling the pie at Helen's Restaurant, a Down East icon.

HELEN'S TOP TEN PIES

Established in 1950 by Larry and Helen Mugnai, Helen's Restaurant in Machias is a Down East icon, famous for its pies. Here are Helen's top ten pies, according to owner Julie Barker.

Apple

Blueberry

Chocolate Cream

Choconut *(aka "the pie of the Gods")*

Custard

Coconut Cream

Strawberry

Graham Cracker

Lemon Meringue

Pumpkin

Cherryfield, the self-proclaimed Blueberry Capital of the World, is home to the factories of the world's largest blueberry growers and processors — Jasper Wyman & Son, a Maine company, and Cherryfield Foods, a Canadian company owned by Oxford Frozen Foods. It is also the gateway to the state's largest stretches of blueberry barrens, sprawling northeast into Deblois, Columbia Falls, Epping, Centerville, and Jonesboro. The town's name comes from the wild cherries that grow along the banks of the Narraguagus River.

The wild blueberries that grow on the barrens of Washington County differ from the familiar marble-size blueberries grown in New Jersey and Michigan in both the way they grow and how they taste. One acre of wild blueberries typically contains well over one hundred varieties of the berry, each one as genetically distinct from the other as a McIntosh apple is from a Delicious. A quart box contains a mix of overripe and underripe fruit along with fruit that's perfectly mature, adding to the complexity of the taste.

ROYAL MONTANA
ORGANIC EGG FARMER AND PROPRIETOR OF THE CHERRYFIELD GENERAL STORE, CHERRYFIELD

"The Cherryfield General Store opened on June 26, 2010. The building was built in 1865 as a boot and shoe factory. When I bought it, it was caving in. In fact, they were working on the road with jackhammers and I thought, 'The whole thing is going to crumble down around me!' It took a year to get it in the condition it's in now. Our motto is "crooked building, straight deals."

"I have a friend with mental retardation, and I wanted a place where people with special needs could learn retail and learn how to interact with the public — and where the public would learn how to interact with them. I also wanted a place where local artists, craftspeople, and farmers could sell their wares. On opening day, we had two people with their consignment. Now we have more than forty-five selling penny candies, wonderful baked goods, yummy pies, chocolates, homemade tablecloths, pillowcases, beauty products, birdfeeders, even hula hoops — all made in this area. Being so close to Route 1, we get a lot of traffic, and it's been very good for getting extra money into people's pockets.

"Cherryfield was a huge lumbering town in the 1800s. The First Union Trust Bank was here in Cherryfield. We had the first electricity in the state — it was called The Little Dynamo and it was in the river and it made electricity. The people coming here brought their money and their taste and their culture, and they built all these beautiful homes.

"I grew up here. When I left, I told myself I was never coming back, that there was no opportunity. But once I left, I couldn't get it out of my mind — how beautiful it is, how wonderful and loving the local community is."

CONNIE HARTER-BAGLEY
CONNIE'S CLAY OF FUNDY, U.S. ROUTE 1, EAST MACHIAS

"I came to Down East Maine thirty years ago to be a hide-in-the-woods hippie. I did homesteading in Trescott, I worked at the research station in West Quoddy, I raked blueberries, I made wreaths — the whole nine yards. I had a lot of fun, but it didn't take me long to realize I better go back to making pottery, that it would be more lucrative.

"I love this spot. It is one of the most beautiful views of any little town in Maine — the cute little river, the cute little bridge, the church steeple. We used to have what we called The Nice Spot Jar. If you said, 'What a nice spot!' you had to put a dollar in the jar. It was more of an idea than a reality.

"When I first came here, I was one of only two potters in Washington County as far as I knew. Now there are lots more artists. They want to be where it's quiet and peaceful, where the air and water are clean. I wouldn't say there is a tremendous amount of support for artists here, though. You need to know how to survive and be frugal and be happy with what you have."

66 1979: Sawed alphabet on pencil first time 99

— ENTRY ON TIMELINE AT RAY MURPHY'S CHAINSAW SAWYER ARTIST LIVE SHOW IN HANCOCK

>>>DON'T MISS

SIGHTS

St. Croix Island International Historic Site, 84 St. Croix Drive, Calais. 207-454-3871. nps.gov/sacr. An interpretive trail with life-size bronze statues of the French explorers who settled here in 1604 and the Passamaquoddy people with whom they traded game and bread.

Moosehorn National Wildlife Refuge, Route 1 between Dennysville and Whiting. 207-454-7161. fws.gov/northeast/moosehorn. Moosehorn's 7,200 acre Edmunds division borders Cobscook Bay. There is a ten-mile network of unmaintained trails.

EVENTS AND ENTERTAINMENT

Machias Wild Blueberry Festival and the **Blackfly Ball,** Machias. 207-255-6665. machiasblueberry.com. Organized by Centre Street Congregational Church, UCC, this five-day festival during the third week of August celebrates all things wild blueberry, the fruit that grows in abundance in this region. The Blackfly Ball, organized by Beehive Design Collective and held on the festival's last night, is a wacky dress-up dance party at Machias Valley Grange Hall and Bad Little Falls Park.

FOOD

Riverside Inn, 622 Main Street, East Machias. 207-255-4134. riversideinn-maine.com. Fine dining alongside the East Machias River, where bald eagles are frequently seen flying overhead. Owner-chef Rocky Rakoczy's signature dish is fresh salmon stuffed with crab and shrimp.

Helen's Restaurant, 111 East Main Street, Machias. 207-255-8423. helensrestaurantmachias.com. A Down East icon, Helen's is best known for its pies. While still offering classic Maine fare, like lobster, fried clams, and fish chowder, Helen's has updated its menu, using locally sourced seafood, meats, cheeses, produce, and dairy.

Le Domaine, Route 1, Hancock. 207-422-3395. ledomaine.com. A fine French restaurant with entrées like beef bourguignon, chou farcie, and coquille St. Jacques en croûte. *The New York Times* pronounced Le Domaine "endearing," and its wine list "outstanding and entirely French."

SHOP

Katie's on the Cove, 635 U. S. Route 1, Robbinston. 207-454-8446. katiesonthecove.com. This brilliant yellow cabin splashed with sixties-style hippie flowers is hard to miss. Offerings include old-time Maine favorites, like Needhams (chocolate, coconut, and potato) and maple pecans, and chocolatiers Lea and Joseph Sullivans' own creations, like mustard truffles made from Raye's Mustard of nearby Eastport.

The Red Sleigh, Route 1, North Perry. 207-853-6688. kendallfarmcottages.com/redsleigh. Crafts and edibles from more than forty local artisans and farmers.

45th Parallel, Route 1, Perry. 207-853-9500. This fun, eclectic gift shop is entirely unexpected on this lonely stretch of Route 1. Everything from t-shirts and sweatshirts to antique beds, even a giant stuffed lion.

Wild Blueberry Land, Route 1, Columbia Falls. 207-483-2583. This giant, grin-provoking blue dome, part amusement park, part specialty shop, seems to have bubbled up from the earth in the heart of the most concentrated collection of blueberry barrens in Maine. Within are freshly made blueberry pies, blueberry jam, blueberry syrup, fresh (in season) and frozen blueberries, and blueberry-inspired books and novelties.

All street names refer to Route 1 unless otherwise noted.

Ellsworth
Orland
Bucksport
Stockton Springs
Searsport
Belfast
Northport
Lincolnville
Camden
Rockport
Rockland
Thomaston

HULL UK

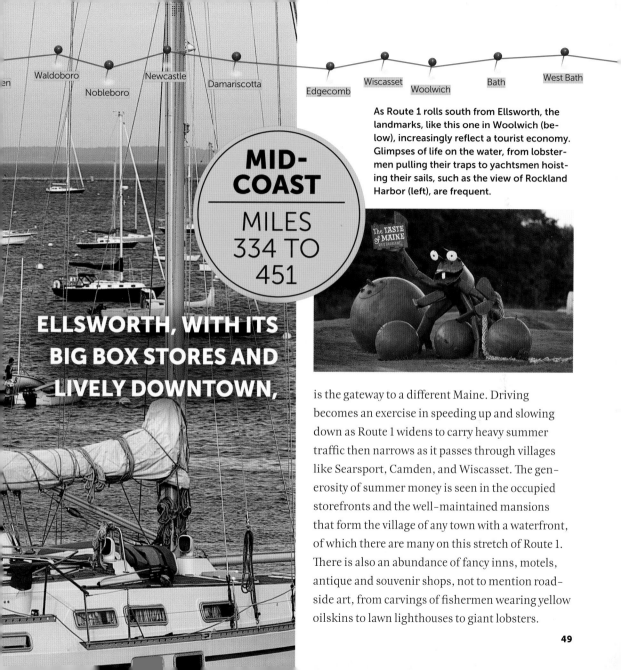

Waldoboro · Newcastle · Damariscotta · Edgecomb · Wiscasset · Woolwich · Bath · West Bath · Nobleboro

MID-COAST

MILES 334 TO 451

As Route 1 rolls south from Ellsworth, the landmarks, like this one in Woolwich (below), increasingly reflect a tourist economy. Glimpses of life on the water, from lobstermen pulling their traps to yachtsmen hoisting their sails, such as the view of Rockland Harbor (left), are frequent.

ELLSWORTH, WITH ITS BIG BOX STORES AND LIVELY DOWNTOWN,

is the gateway to a different Maine. Driving becomes an exercise in speeding up and slowing down as Route 1 widens to carry heavy summer traffic then narrows as it passes through villages like Searsport, Camden, and Wiscasset. The generosity of summer money is seen in the occupied storefronts and the well-maintained mansions that form the village of any town with a waterfront, of which there are many on this stretch of Route 1. There is also an abundance of fancy inns, motels, antique and souvenir shops, not to mention roadside art, from carvings of fishermen wearing yellow oilskins to lawn lighthouses to giant lobsters.

JIM PENDERGIST
OWNER WITH MARK ROSBOROUGH OF 112 MAIN STREET, ELLSWORTH

"I came to the area in 1966. I was stationed with the Navy in Winter Harbor, and I stayed. Ellsworth has changed a lot, especially High Street (Route 1, east of downtown) and up on Beckwith's Hill, where Home Depot and Wal-mart and all that development is. There was nothing back then. When my wife Brenda was a kid here, there were fields going out High Street.

"Downtown has maintained its integrity. It hasn't changed much. Businesses come and go — the old Watson's Dry Cleaner is now Simone's restaurant, Western Auto is now H&R Block — but you haven't seen the buildings replaced or new buildings going up. This building is one of only two on Main Street that is still close to its original condition; the other is The

Grand theater. The Luchini family built this place in 1933 after a fire destroyed everything on Main Street. They had a restaurant that was a long-, long-time success.

"Summer was always a bustling time, and still is. I used to pump gas at the High Street Sunoco and traffic was backed up over the hill, with only one way coming in and one way going out. Ellsworth is a retail hub for Hancock and Washington counties. The population is just over 7,000, but we've got fifty to sixty thousand people coming to Wal-mart, going to Home Depot, going to these big shopping places and the restaurants. We are not wholly dependent on the tourists for our economic success."

Searsport is home to a number of flea markets, antiques shops, auction houses, used bookstores, and junk shops. Left, Finn's Irish Pub serves thirsty locals.

The Penobscot Narrows Bridge, linking the towns of Prospect and Verona Island, is arguably Maine's most beautiful bridge. Completed in 2006, it is one of just three cable-stayed structures in the United States with a cradle system whose stays are carried from the bridge deck through the tower and back to the bridge deck as a continuous element (the others are the Zakim Bridge in Boston and the Veterans Glass City Skyway in Toledo).

The bridge's dramatic design softened the loss of the span it replaced, the Waldo-Hancock Bridge, a beauty and engineering marvel in its own right. Designed and built by Robinson & Steinman of New York, it made news from the start. It was Maine's first long-span suspension bridge, and it was named The Most Beautiful Bridge Constructed in the United States in 1931 by a committee of architects and engineers appointed by the American Institute of Steel Construction. Robinson & Steinman used innovative technology, including pre-stressed twisted wire strand cables and Vierendeel trusses, whose members form rectangular rather than the conventional triangular openings. Given the bridge's historic and architectural credentials, it's no wonder that news of its irreparable structural problems was greeted with public dismay in 2003.

The two bridges stood side by side for more than six years, the new one safely carrying cars and trucks across the Penobscot Narrows, the old one closed and deteriorating. The Waldo-Hancock Bridge was finally demolished in the winter of 2013.

What did Colonel Jonathan Buck do to deserve being cast as the cruel villain in one of Maine's most enduring ghost stories? Apparently nothing. The founder of Bucksport, who died March 18, 1795, is buried under a monument bearing a stubborn leg-shaped stain, and through no fault of his own, his reputation has been stained in kind. The granite obelisk stands at the front of Buck Cemetery, facing Route 1 in Bucksport.

The tale inspired by the gravestone discoloration asserts that Buck condemned a woman for witchcraft and ordered her to death by burning. As the sentence was being carried out, the woman cursed the colonel, saying her sign would be forever upon his tombstone. As the flames consumed the woman's body, her leg fell away and rolled out of the fire. The monument's leg-shaped stain, the story concludes, is the woman's curse realized, and all efforts to scrub it away have failed.

The legend has been so tenacious — even Robert P. Tristam Coffin retold it, thinly disguising the colonel's identity by changing the first letter of his name, in his poem "The Foot of Tucksport" — that a sign has been posted next to the monument in an attempt to set the record straight. The sign reads in part, "There is no record of anyone being executed for witchcraft in Maine. Stories

that the monument has been replaced are untrue. This is the original. Stonecutters say it is not unusual for granite to contain a flaw such as a stain, which appears only after cutting and polishing. The outline can be removed but reappears when air oxidizes the iron. … The facts surrounding the life of Colonel Buck are that he was an honorable and industrious man who founded this community and was a leader in its early development, building the first sawmill, the first gristmill, and the first boat. Notably the witch's curse was unheard of before the flaw in the monument appeared."

A maritime museum, roadside fishmongers, blueberry processors, even llamas — Route 1 in midcoast Maine is distinguished by its incredible variety.

KIM O'BRIEN
MANAGER OF PERRY'S NUT HOUSE, 45 SEARSPORT AVENUE, BELFAST

"About nine years ago, I was living in Illinois, a single mother with an eleven-year-old daughter, working fifty-five to seventy-five hours a week as an HR manager for Home Depot. I hated that job. My parents, who had retired to Belfast, thought they could help by buying Perry's Nut House and having me run it for them. We had about three weeks to get it ready for opening on Memorial Day Weekend of 2004.

"None of us was familiar with Perry's history, but I hired all local people to work for me and they told me about it. Customers started sharing their memories, too. It wasn't long before I threw out my idea of what the business was going to be.

"The founder of this store is I.L. Perry, who originally had a cigar store here. In 1925, there was a bumper crop of pecans in Georgia. Perry had some shipped to Belfast, where he roasted and salted them and sold them out of a downtown storefront. They sold so well, he moved the nut operation over here. Joshua Treat III bought the store at age nineteen in the 1940s and added oddities

like the museum of taxidermy mounts that people remember. All of those animals were sold at auction by another owner in 1997.

"Slowly over the years we've added back the items people remember from their childhood: gag gifts, stink bombs, whoopee cushions, fake dog poop, and fake vomit, and old-fashioned toys like metal spin tops, train whistles, jacks, tiddley winks, and pickup sticks. We're selling handmade wooden toys made by Ed Parent, the Toymaker of Maine — things like a wooden toaster with wooden toast, a wooden chain saw.

"A lot of people think we're actively seeking to bring the taxidermy mounts back to Perry's Nut House, but that's not quite true. A woman who bought some pieces at auction was moving and gave us an opportunity to bring them home. I bought them because they are the iconic Perry's pieces that everyone remembers: the gorilla, the python, which is 21 feet long, and the alligator, which is 16 feet long. Someone else offered me the lion that used to be here, but nobody remembers the lion, so I said no. In this day and age, when it comes to food handling and sanitation — we sell a lot of fudge and snack mixes — we want to keep the animal presence to a manageable level!"

Neither rain, nor snow, nor sleet, nor hail shall keep the postmen from their appointed rounds. But twenty-foot-high mailboxes? That's another matter. The mailboxes, one labeled BILLS, the other AIRMAIL, have defied mailmen in East Belfast for decades. Or so owner Jane Horne assumes — she's never climbed a ladder to check the boxes for mail. Her husband, the late Allen Horne, installed the mailboxes for no reason other than it struck him as funny. "He was always doing things like that," she says, with a chuckle.

Adultery! Lies! Murder! Camden!

Nearly sixty years ago Camden was Peyton Place. We're not being metaphorical here. Camden really was Peyton Place for the month of June in 1957, when it served as the setting for Twentieth Century Fox's movie version of *Peyton Place,* the controversial novel by Grace Metalious that portrayed life in a small New England town. "The town was full of movie stars, actors, film crews, and the air was electric with excitement," says Terry Brégy, who narrated a trolley tour of the movie's landmarks during the town's fiftieth anniversary celebration of the film in 2007. "And after the movie came out, the town was changed in many ways — some good, some not so good, depending on who you ask."

Although it is tame by today's standards, Metalious's novel, which exposes the moral hypocrisy hiding behind Peyton Place's tranquil, prim façade, was scandalous when it was published in 1956. Indeed, many public libraries, including Camden's, banished it from their shelves, yet it was wickedly popular, selling 12 million copies.

Twentieth Century Fox chose Camden for the setting after failing to find suitable locations in Vermont and New Hampshire. Five hundred locals were hired as extras. The movie, which starred Lana Turner (whose Camden scenes were played by a double), Hope Lange, Lee Philips, and Arthur Kennedy, had its world premiere at the Camden Movie Theater on December 11, 1957, a big deal for a small mill and summer resort town.

The buildings and landmarks seen in the movie haven't changed much. Among them is First Congregational Church, which appeared in a montage of churches.

"The director wanted to underscore the irony that a town with so many sordid secrets would have so many fine churches," Brégy says. "This raised a few eyebrows amongst the local parishioners."

The Town of Rockport has made preserving the spectacular view of Clam Cove from Route 1 a priority.

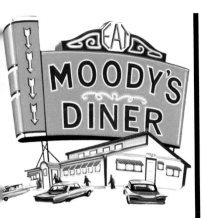

ROAD FOOD - ICONIC EATERIES

Rock's Family Diner, Fort Kent

Dolly's, Frenchville

Helen's, Machias

Chester Pike's Galley, Sullivan

Cappy's, Camden

Wasses Hot Dogs, Rockland

Moody's, Waldoboro

Red's Eats, Wiscasset

Miss Brunswick, Brunswick

Congdon's Doughnuts, Wells

Maine Diner, Wells

Ogunquit Lobster Pound, Ogunquit

Flo's, York

Bob's Clam Shack, Kittery

Warren's Lobster House, Kittery

"Perry Greene was the P.T. Barnum of his time. He had a huge presence and a booming voice. In 1941, he and his stepson took a team of seven Chinook dogs on the longest sled dog trek ever made in the lower 48: They hauled eight hundred pounds of equipment from Fort Kent to Kittery — 502 miles in ninety hours. The *Saturday Evening Post* did several articles on the trip, and Chinooks were elevated to absolute legend. Dorothy Lamour met Greene in Portland and took a Chinook. Governor Frederick Payne met him and got a Chinook, Dirigo. Thousands and thousands of people visited here — Perry Greene and his wife Honey built this log home and kennel in 1947. The Fort Kent-to-Kittery team is buried here.

"All Chinooks trace their bloodline to Arthur Walden's lead sled dog, Chinook, who was bred from a mastiff stray and a descendant of Admiral Peary's lead dog, Polaris. Greene acquired the Chinook breed in 1940, and was very possessive of it. People who wanted to purchase a Chinook had to stay overnight and meet

his dogs. He had an exclusive on the breed; he only let spayed females leave the kennel. After he died, there was no clear successor, and Chinooks fell on hard times. By 1981, there were only twelve breedable dogs. A group of breeders, including me, divided the dogs up all over the country so we could begin bringing back the breed. Today, there are probably six hundred to eight hundred Chinooks. There is a national organization of Chinook owners and the breed is recognized by the United Kennel Club.

"My Uncle Stan Victor got his first Chinook from Perry Greene in the early sixties, so Chinooks have been in my family for a long time. At one time, I had twenty-seven Chinooks, and I am still breeding them. In 1993 my wife, Martha Kalina, bought this house and kennel, which celebrated its sixty-fifth anniversary in 2012."

Gigantic lobster rolls have made Red's Eats world famous, but some people blame the takeout stand's popularity for Wiscasset's summertime traffic jams.

Wiscasset's collection of nineteenth-century sea captains' mansions have earned it bragging rights to the motto "Prettiest Village in Maine," but did you know that Wiscasset also is the Worm Capital of the World?

Bloodworms, so named for their rusty hue, and blue-headed sand-worms, have been harvested in the Sheepscot Estuary's mud-flats for more than forty years. The writhing, foot-long creatures, which can deliver a stinging bite, are used as bait by salt water fishermen.

If you've driven through Wiscasset in the summer, you've seen it: the long line of people trailing from Red's Eats, the little red takeout stand at the corner of Route 1 and Water Street. What are they waiting for? Lobster rolls. Red's offers a simple, but gargantuan, version of the sandwich that has attracted mountains of publicity from food and travel writers, television chefs, and morning show hosts. It is the creation of the late Al Gagnon, who owned Red's Eats for more than thirty years and whose family continues to run the seasonal operation today.

The concession's roots stretch even deeper into Route 1 history, to at least 1938, when Leland and Mabel Bryant wheeled a food wagon to the spot, christened it The Little Trailer, and began selling sandwiches and soft drinks. In 1950, after several changes in ownership and names, nineteen-year-old Alan Pease bought the wagon and named it Al's Eats. It helped pay the future Maine District Court chief justice's way through law school. Pease eventually replaced the food wagon with the little shed that houses Red's Eats today.

The original Red was a later owner, Harold Delano, who possessed a head of ginger locks.

The stand went through several more changes of ownership before Al Gagnon bought it in 1977 — and inherited Delano's nickname in the process. After sampling a disappointing lobster roll at another restaurant, Gagnon put his mammoth version of the sandwich on the menu and a Maine icon was born.

Next time you're stuck in traffic watching the tourists queue up at Red's, you may want to pay your respects to the Wiscasset bypass. After more than fifty years of studies and heated debate, the plan to re-route Route 1 around Wiscasset village, which in summer produces one of the most congested stretches of roadway in the state, is dead. Maine Transportation Commissioner David Bernhardt pulled the plug in 2011, saying, "At some point you have to say, 'Enough is enough.' "

It wasn't entirely unexpected. In 2010, the United States Army Corps of Engineers selected a potential bypass route from one of three northern alternatives — the farthest the concept had ever advanced. A month later, a bald eagle's nest, which is protected by federal law, was found. The Maine Department of Transportation was forced to abandon the route. It had no viable alternatives.

Ultimately, the decision to abandon the Wiscasset bypass came down to money. "Due to the excessive cost and limited benefits, the only sensible choice was to cancel the plans to build a bypass," Bernhardt said.

Bath Iron Works' giant cranes can be seen from the Sagadahoc Bridge, which carries Route 1 over the Kennebec River from Woolwich to Bath.

>>> **DON'T MISS**

SIGHTS

Penobscot Marine Museum, Route 1, Searsport. 207-548-2529. penobscotmarinemuseum.org. Spread throughout the campus of ten historic buildings are a display of New England's largest collections of small boats, an exhibit about Maine fisheries, ship models, folk art, and more.

Farnsworth Art Museum and Wyeth Center, 16 Museum Street (entrance just off Route 1), Rockland. 207-596-6457. farnsworthmuseum.org. The Farnsworth is best known for its collection of works by the famous Wyeths — Andrew, N.C., and Jamie. Its holdings also include other well known artists with Maine connections, such as Marsden Hartley, Rockwell Kent, Neil Welliver, Alex Katz, and Robert Indiana.

EVENTS AND ENTERTAINMENT

Camden Windjammer Festival, Camden. camden-windjammerfestival.com The fifteen schooners that comprise the windjammer fleet sail into Camden Harbor on Labor Day weekend, when the waterfront is jammed with thousands of visitors. The schedule includes concerts, boat building demonstrations, craft shows, and a lobster crate race, in which participants must run across a string of floating lobster crates.

Maine Lobster Festival, Rockland. This a loud, steamy, jubilant affair that draws nearly 100,000 people to Rockland in early August. Most come for the lobster — some 20,000 pounds of the steamed crustacean are served over the course of the five-day celebration.

RECREATION

Camden Hills State Park, 280 Belfast Road, Camden. The ledgy, open summits of Mounts Battie and Megunticook are among the striking features of this 6,500-acre park. There is a roadway to the top of Battie, which overlooks Camden Harbor and Penobscot Bay. There are nineteen hiking trails and a graded five-mile path through the woods.

Windjammer fleet, Camden Harbor. Several schooners call Camden home and offer cruises ranging from a couple of hours to several days spring through fall. Just amble down to the boardwalk and check out the offerings.

FOOD

Cleonice. 192 Main Street, Ellsworth. 207-664-7554. cleonice.com. Creative cuisine with an emphasis on Mediterranean dishes, like bouillabaisse de Marseille, paella, and pork ribs and sausage ragu.

Boyton-McKay Food Co., 30 Main Street, Camden. 207-236-2465. boynton-mckay.com/our-menus. Charming former drug store and soda fountain serves inventive hot and cold wraps, burritos, stir-fries, salads — and breakfast all day.

Thomaston Café and Bakery, 154 Main Street, Thomaston. 207-354-8589. thomastoncafe.com. Chef-owner Herbert Peters' menus are abundant in locally grown fruits and vegetables, breads and pastries baked on site, and simply prepared, perfectly cooked Maine seafood. Even the butter is fresh.

Moody's Diner, Route 1, Waldoboro. 207-832-7785. moodysdiner.com. A Maine icon, this 1930s diner is famous for its classic diner fare like corned beef hash, hot turkey sandwiches, fish and chips, and pies.

SHOP

Big Chicken Barn Books & Antiques, Route 1, Ellsworth. 207-667-7308. bigchickenbarn.com. An enormous (21,600 square feet) former chicken coop is packed with antiques and what is proclaimed to be the largest selection of books by Maine authors.

Swan's Island, 231 Atlantic Highway, Northport. 207-338-9691. swansislandblankets.com. Formerly Swans Island Blankets, this little textile company now makes wool and wraps in addition to blankets, and it sells spun and hand-dyed yarns.

Cellardoor Winery at the Villa, Route 1, Rockport. 207-763-4478. mainewine.com. Located in a handsome yellow Victorian house, Cellardoor offers guests complimentary wine tastings and, on Saturdays in summer, food and wine pairings. Cellardoor's vineyard and main store and tasting room are in nearby Lincolnville.

Maine State Prison Showroom, 358 Main Street, Thomaston. 207-354-9237. Handcrafted by prisoners, the hundreds of wooden products include boat models, dressers, coffee tables, spice racks, birdfeeders, and more.

Fawcett's Antique Toy & Art Museum, Route 1, Waldoboro. 207-832-7398. home.gwi. net/~fawcetoy. Enormous collection of antique toys and original comic art, including Mickey Mouse and Lone Ranger memorabilia.

All street names refer to Route 1 unless otherwise noted.

Brunswick

Freeport

Yarmouth

Cumberland

Falmouth

Portland

South Portland

Scarborough

Saco

Biddeford

Arundel

Kennebunk

Wells

Ogunquit

York

Kittery

Route 1 parallels the Androscoggin River in Brunswick (left). A bicyclist crosses the Swinging Bridge, which spans the Androscoggin between Brunswick and Topsham (below).

SOUTHERN MAINE

MILES 452 TO 527

RUNNING PARALLEL TO INTERSTATE 295 FROM BRUNSWICK

to South Portland and then alongside Interstate 95 until it reaches the New Hampshire border in Kittery, Route 1 is now the slow road. Indeed, in some places, such as Brunswick and Yarmouth, Route 1 seems to want to rid itself of traffic as it curves sharply east, fooling drivers who aren't paying close attention into driving straight onto the Interstate. And in Portland, which seems like a major metropolis after nearly five hundred miles of small towns, farmland, and wilderness, Route 1 and I–295 are the same road.

With the exception of a few miles here and there, these last seventy-five miles of Route 1 are a string of fast-food restaurants, car dealerships, and big-box stores. It is not a pretty road, but it is an essential one, the lifeblood of the communities that have grown up around it.

This 1912 postcard shows Brunswick Falls on the Androscoggin River, which powered the Cabot Manufacturing Company, a textile mill.

PHILIP WAGNER
OWNER,
DEROSIER'S
MARKET,
120 MAIN STREET,
FREEPORT

"My great-great-grandfather, Augustus Derosier, opened Derosier's Market in 1904, so we were here eight years before L.L. Bean. Since then, their sales have broken the $2 billion mark; ours are not quite there yet! We've had multiple offers to sell; I always say the business has supported five generations, so any offers would have to support the next five, which tends to thin the herd very quickly.

"In 1904, it was all grocery. My grandmother, Alice Wagner, introduced pizzas and sandwiches sometime in the fifties. When I was a kid — I was born in 1969 — there were multiple shoe factories in town, and noontime was chaos. The workers would line up all the way around the corner for their sandwiches. I remember the wall of sandwiches on the counter. We made ham Italians, and you could get them either with onions or without. Those were the options. There was no light this or extra that.

"I worked here when I was a teenager, and I couldn't wait to get away from my hometown. So I went off and did other things. I worked for Senator Richard Cohen until he went to Washington on me. After that I ran the recreation department here in town and in Falmouth. One day in 2001 I quit my job in a huff, and I asked my dad if I could work for him here for the summer. He said, "I'm ready to go if you want it."

"Under my dad's ownership, it was more of a convenience store than a grocery. Now every gas station is a convenience store, so we've adjusted again — we're more food, more restaurant. We get a lot of tourists in summer, but the locals are my bread and butter even though there are a lot fewer of them living downtown — all that used to be housing is now parking lots. When people who were born and raised in Freeport come back for a visit, they always come here because they don't recognize anything else in town."

It seems they like things oversized in Freeport: The shopping mecca is home to both L.L. Bean flagship store, whose entryway is marked by a giant hunting shoe, and mapmaker DeLorme, whose headquarters house Eartha, the world's largest rotating and revolving globe.

If Maine held a contest for Most Changed Town, Freeport would be a strong contender. Thirty-five years ago it was a sleepy little village, and with the exception of L.L. Bean, its handful of shops were focused on serving locals. Then, in 1981 a fire destroyed an apartment block across Main Street from L.L. Bean. A new building was erected, and Freeport's first national retailer, Dansk, moved in. More factory outlet stores soon followed, transforming the town into one of New England's premiere shopping destinations.

A forty-foot-tall Indian in full headdress has been welcoming people to Freeport since 1969, when Julian and Bill Leslie had him installed to lure customers into their Casco Bay Trading Post, which sold leather and sheepskin goods, hand-sewn moccasins, and souvenirs. The Leslies named him Chief Passamaquoddy, but most people know him as The Big Indian or the BFI (you figure it out).

Made of fiberglass, plywood, and steel rods, The Big Indian is the work of Pennsylvania artist Rodman Shutt, who has made dozens of roadside statues, including Boothbay Harbor's giant Old Lobster Fisherman. Shutt's Charlemont Indian in Charlemont, Massachusetts, who could be the BFI's little brother — he's only twenty feet tall, but he bears a striking resemblance. According to some published accounts, the Big Indian was such a distraction when it was shipped to Maine aboard a flatbed truck that part of the New Jersey Turnpike was shut down.

The Casco Bay Trading Post closed years ago, and in its place have been a succession of businesses, including Levinsky's, American Skiing Company, Winter People, and today, Conundrum Wine Bistro. The Big Indian has weathered all the changes handsomely. You might say he's a standup guy.

The five-story brick **B&M** plant in Portland's East Deering neighborhood has been making baked beans on the shores of Casco Bay since **1927** (the factory itself was built in 1913, when B&M made codfish cakes and fish flakes). Founded by George Burnham and Charles S. Morrill, B&M is now owned by New Jersey-based food giant B&G, whose many brands include Cream of Wheat, Underwood, and Ortega, but it still makes beans the way it always has, in enormous cast-iron pots inside brick ovens. On some days, drivers on Route 1 (merged here with I-295) are treated to the fragrance of molasses and sugar as they drive past.

Over the last decade, Portland has come into its own as a dining destination. Writes Julia Moskin of The New York Times, "Portland has undergone a controlled fermentation for culinary ideas — combining young chefs in a hard climate with few rules, no European tradition to answer to, and relatively low economic pressure — and has become one of the best places to eat in the Northeast."

Fitzpatrick Stadium, which sits alongside Route 1 in Portland, is the site of one of Maine's oldest football traditions, the Thanksgiving Day matchup of cross-town rivals Deering and Portland High schools. The Rams (Deering) and the Bulldogs (Portland) played their first Turkey Day game in 1911 and have missed only one since — that was in 1920 due to bad weather. The now defunct Bayside Park was the original setting for the game. In 1931, the event was moved to 6,300-seat Fitzpatrick, where it has been played ever since. Although Deering won the hundredth Thanksgiving game last November with a score of 33-0, Portland can take consolation in the fact that it has the historical edge, having won fifty-four of the standoffs to Deering's thirty-nine (seven games ended as ties).

LINDA WOODARD
DIRECTOR, SCARBOROUGH MARSH AUDUBON CENTER, SCARBOROUGH

"At 3,100 acres, Scarborough Marsh is the largest saltmarsh in Maine. Saltmarshes are formed by barrier beaches, and since pretty much everything north of Portland is rocky coast, there is no other place like it in the state. Scarborough Marsh also is special because it has been preserved. Forty years ago, Dick Parks of Maine Department of Inland Fisheries and Wildlife, and Dick Anderson, who was the executive director Maine Audubon at the time, had a lot of foresight. They went around and figured out who owned all the little pieces of it and they pulled it all altogether. It is now owned and managed by IF&W.

"The marsh is an important stopover for migratory birds flying north to breed. It also is habitat for some endangered species like roseate and least terns, who nest off the coast on Stratton Island and come to the mouth of the marsh to feed. Nelson's and saltmarsh sparrows breed in the marsh. It is a nursery — seventy percent of the fish species we eat depend on the marsh as either a place to nest or spawn or feed. And it's important for flood control. A lot of places have filled in their marshes and paid the price — look at New Orleans — but a marsh as big as Scarborough Marsh can absorb a lot of water. And as the water goes through the marsh, it filters it so it holds any pollutants in the mud and sand.

"One of the things that is fun about working at the Scarborough Marsh nature center is that so many people don't plan their visit — they stop in because they're driving by and they see our canoes and wonder what is going on. I've seen that ah-ha moment, where they realize, wow, this is beautiful. I've been working at the marsh for nearly twenty-five years, and I still find it to be an awe-inspiring place. Every year there are surprises, but I also love the familiarity, the rhythms of the seasons, seeing the same birds come back year after year."

Christopher Hersey holds his nose as he slides down a ride at Funtown Splashtown USA in Saco.

IRA ROSENBERG

PHILANTHROPIST AND OWNER OF PRIME MOTOR GROUP, WHICH HAS SIX CAR DEALERSHIPS IN SACO AND ONE IN SCARBOROUGH, ALL ON ROUTE 1

"I was one of those kids who joined the Navy the day they turn seventeen. I went in to catch the end of the Korean G.I. Bill, and when I came out I worked part-time changing tires for a Chevrolet dealer in Cambridge, Massachusetts, while I went to school. After a year and a half of college, I couldn't stand it, so I quit and went to work full time, first in the service department, then in sales at a Cadillac dealership in Lynn. With $140, I bought a used car operation in Salem. I stayed with that for seven years. Then I bought a Toyota store in Danvers that was going bust. I put it out on the highway and built several other dealerships after that. I had seven or eight dealerships in Danvers alone.

"About ten years ago, my wife, Judith, got sick so we decided to retire to Florida. After four years, I told her, 'Either give me a lobotomy or let me go back to work.' (She opted for the lobotomy!) So I bought this little Toyota store in Saco in 2004. It was a small store selling very few cars, and we took it over and grew it like crazy. We took back our market share and whatever else was available. Four or five years later we built a brand new dealership across the street. It is one of the most exciting dealerships in Maine. I've put in features other dealerships don't have, like a full café, a pool table, and a fish tank that kids can put their hands in to touch the turtles and fishes.

"I believe in taking care of customers personally. My whole life is on the floor and talking to customers. I love doing the television commercials. The camera doesn't scare me. I'm the same person on camera as I am off camera. I did commercials when I was in Massachusetts, but not as much as I have here. I was younger with a family growing up and I thought differently then. Now money isn't my overall reason for being in business. I'm here because I want to have a good time.

"My father was too poor to be a philanthropist, but early on I became friendly with some people who became successful and they taught me philan-thropy. My wife and I have become involved in many different charities. Sweetser is one of my favorites because they do so much for children. I believe I have to continue giving because if I don't the guy upstairs is going to take everything away from me. I'll tell you the truth: the more you give, the more you get. You feel that good stuff in your heart."

The Saco River ambles more than eighty miles through Maine before it reaches the York County cities of Saco and Biddeford. The sister communities have strikingly different personalities, but their fates have been intertwined for two hundred years: Biddeford is a gritty blue-collar town, where 12,000 people, many of them Franco-American, once labored in the mills; white-collar Saco is where the mill owners once lived. For the last decade or so, both cities have been focused on renewal. The shuttered mills have been redeveloped and are occupied by an ever-growing population of artists and entrepreneurs. In late 2012, one of the biggest obstacles to revitalization, a smelly trash incinerator in Biddeford's downtown, closed, raising hopes for a new era of economic development and vibrancy.

Nikki Hunt performs at Bentley's Saloon, a biker bar in Arundel. The establishment is filled with motorcycles (they even hang from the ceiling) and the racecar memorabilia of its owner, retired racecar driver Bentley Warren.

To travelers, Wells is "The Friendliest Town in Maine," a place whose spirit falls somewhere between the boutique-y Kennebunks and honky-tonk Old Orchard Beach. Route 1 and the beaches define a roughly seven-mile-long, two-mile-wide vacation zone crammed with 2,000 motel and hotel rooms, 3,300 RV and campground spaces, and 2,000 seasonal homes. Heavily developed, it is nonetheless uniquely, some might say charmingly, Wells. Route 1 offers frequent glimpses of marshes in the Rachel Carson National Wildlife Refuge, even a few ocean vistas. It is the address for scores of seasonal busnesses, but also groceries, schools, the public library, and the police and fire departments. The vast majority of comercial enterprises are one of a kind, homegrown places like Congdon's Doughnuts, Mike's Clam Shack, and the landmark Maine Diner. National chains are few, in part due to an ordinance prohibiting big-box stores.

VINCE BRAZEN
WHEELS N WAVES, 579 POST ROAD, WELLS

"People started surfing around here sometime in the late sixties. The most common places to surf are the Ogunquit River mouth and the Wells jetty, with these big beaches and parking, there are a lot of good places. With Google maps and everything else, there are no secret spots anymore.

"Interest has exploded in the last six to eight years. Surfing is cool again. Hollywood helps. They made the movie last year about Bethany Hamilton, the girl who lost her arm in a shark attack, and we saw a little bump of interest among young girls. The scene here used to be guy-dominated, but now it's getting close to fifty-fifty. Another reason surfing has become popular again is that while it's not the easiest thing in the world, it's not the hardest thing either. The more you put into it, the more you get out of it, and most of the local surf shops offer lessons.

"We're the oldest surf shop in Maine by a mile, but all shops have been here for a while. We're all real civil — we're not calling each other to go out to dinner, but we get along because we have something in common. The local surfing community, from York to Kennebunk, is very tight.

We lost a good friend from the surfing community in a car accident a few months back. We did a traditional Hawaiian paddle out, where we all paddled out and threw flowers and gave some speeches. You really get to see the love of the community with something like that."

Big Hammer, Central Building Supplies, Madawaska

Shoe Tree, Hodgdon

Abandoned Esso Station, Waite

Log People Village, Perry

Wild Blueberry Land, Columbia Falls

The Man in a Crib on a Roof, Linwood's, Gouldsboro

The Real Chainsaw Artist, Hancock

Stairway to Nowhere, Maine Spiral Stairs, Searsport

Air Mail Mailbox, Belfast

Dalmation Gondola, Bath Cycle and Ski, Woolwich

Osprey Nest on Median Strip, Bath

Cow in a Coffee Cup, Sisa's BBQ, Brunswick

The Big Indian, Freeport

Moose Crossing Sign, Kennebunk Village

Warren's Lobster House Billboard, Kittery

Sandwiched between Routes 1 and 1A, York's Wild Kingdom is part zoo, part amusement park. Riders on the Ferris wheel can see York's sandy beaches just a couple of blocks away. One of just two zoos in Maine (the other is Acadia Zoological Park on Route 3 in Trenton), York's Wild Kingdom displays more than 75 exotic animals, from white-bearded wildebeests and Bengal tigers to tarantulas and colorful butterflies.

A girl enjoys a lobster-shaped lollypop at Stonewall Kitchen's flagship store in York.

Well known for its beaches, the town of York also is a boating destination. The mouth of the York River, just east of Route 1, provides a perfect inlet for yachts to moor. In summer, York Harbor is filled with Hinckleys, Hatterases, and other fine examples of floating money pits.

Nearby is the historic Cliff Walk, which follows the Eastern Point Ledges from Harbor Beach to Cow Beach. In recent years, the pathway has been the subject of controversy as the town waged a legal challenge against two oceanfront mansion owners who erected fences blocking public access.

Fisherman's Walk, by contrast, is open to walkers. The path skirts the York River, whose personality changes with the tides, and leads to the Wiggly Bridge, a small pedestrian suspension bridge to Steedman Woods, a public preserve with a loop trail along the estuary.

The very last (or very first business on Route 1 in Maine is Back Channel Canvas Shop. The maker of boat covers, awnings, and other things canvas is perched on the southern bank of Badgers Island in Kittery. Its big picture windows face Portsmouth, New Hampshire, on the opposite shore of the Piscataqua River.

The Piscataqua offers plenty of entertainment. "A lot of traffic goes up and down that river," says Guy Sinclair, an employee of the canvas shop. "There is a steady stream of fishing boats, gypsum ships, salt ships, and oil barges." Tourists often stop to snap photographs of their entry into Maine.

Badgers Island took on a new personality when the rusty eighty-nine-year-old Memorial Bridge was closed to vehicle traffic in 2011. Route 1, named Island Avenue here, became a dead-end street overnight. "I have to admit I kind of like the quiet," says one resident. For a while, the shuttered bridge became a favorite walking and bicycling route for area residents, but then span was demolished in the winter of 2012. A new Memorial Bridge is scheduled to be completed by summer 2013.

Badgers Island, located in the Piscataqua River, carries Route 1 between Maine and New Hampshire. It is connected to the Kittery mainland by the Badgers Island Bridge, and to New Hampshire by the new Memorial Bridge, which is slated to open in the summer of 2013.

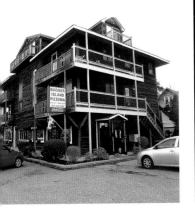

PETER IORDANOU
OWNER, BADGERS ISLAND PIZZERIA, 3 ISLAND AVENUE, KITTERY (MILE 527)

"I've been in this part of the country for forty-one years. I grew up over the bridge in Portsmouth. I live in New Hampshire now, but I lived in this building for eight or nine years. They call this the seacoast area, and it is a unique place. It's safe, and the people are better here than any place that I've traveled to in America.

"We're only two hundred yards from the coast. We're connected by bridges to the mainland, but you still feel like you're on an island. You see seabirds and, except for Route 1, the island is quiet and secluded. It's mostly residential, but

there are a couple of big businesses, Greenpages, a technology business, and Weathervane, which has its lobster warehouse here. Lately people are putting really expensive luxury apartments here. Portsmouth is the only port in New Hampshire, so there are big boats going through drawbridges all the time. Portsmouth and Kittery were established in 1653. The communities are very close — it's like one town."

>>> DON'T MISS

Fishway Viewing Room, Brunswick–Topsham Hydro Station next to Fort Andross, Maine Street (just off Route 1), Brunswick. From mid–May to June, spectators can watch alewives, salmon, eels, and other fish climb forty-two watery steps to cross the dam and return to their spawning grounds in the upper Androscoggin River.

Funtown/Splashtown USA, Route 1, Saco. 207-284-5139. funtownsplashtownusa.com. An amusement and water park with giant water slides, thrill rides, kiddie rides, and games.

Brick Store Museum, 117 Main Street, Kennebunk. 207-985-4802. storemuseum.org. Located in four linked buildings, the museum focuses on the Kennebunks' cultural heritage with permanent and changing exhibits.

EVENTS AND ENTERTAINMENT

Summer in the Park, L.L. Bean Discovery Park, 95 Main Street, Freeport. 800-441-5713. llbean.com/summer. Free concerts and events such as food festivals, dog contests, and games take place on the L.L. Bean campus all summer long.

Ogunquit Playhouse, 10 Main Street, Ogunquit. 207-646-5511. Ogunquitplayhouse.org. An iconic summer stock theater offering live, professional productions from June to October. The roster of famous actors who have performed here since it opened in 1950 includes Bette Davis, Art Carney, Steve McQueen, and Jessica Tandy.

La Kermesse Franco-Americaine, downtown Saco and Biddeford. lakermessefestival.com. Founded in 1982, this three-day mid–June Franco-American celebration has recently expanded to include a world stage with international music and dance performances. Naturally, there's an emphasis on things francais, including a crepe breakfast and Acadian bands.

RECREATION

Androscoggin River Bicycle Path, lower Water Street to Grover Lane, Brunswick. This paved bikeway stretches 2.6 miles along the Androscoggin and parallel to Route 1.

Back Cove Trail, just off Routes 1/295, Portland. Popular with walkers, runners, bicyclists, and dog walkers, this 3.5-mile loop around Back Cove offers great views of the city skyline. And since you've detoured off Route 1 already, you really must go downtown and check out the boutiques and restaurants in the Old Port and other neighborhoods on the Portland peninsula.

FOOD

Ice It! Bakery, 305 Route 1, Yarmouth. 207-847-3305. Customers decorate cupcakes and mini-cakes with the brightly colored sugars and sprinkles in the magnetized pots that dot the wall. It's fun, and, better still, the rich and dense cupcakes are superb.

Maine Diner, 2265 Post Road, Wells. 207-656-4441. mainediner.com. Classic diner fare like meatloaf, shepherd's pie, Yankee pot roast, and New England boiled dinner. Signature dishes include seafood chowder and lobster pie.

Flo's Hot Dogs, 1359 Route 1, Cape Neddick, floshotdogs.com. Customers routinely wait thirty minutes for these steamed weiners, which by many travel writers' accounts are pretty ordinary. The allure is the hot sauce. "Dark and a bit chunky, it gently teases between sweet and sour," the *Washington Post* writes. "It's a bit relishy, vaguely chutney-like and altogether mysterious."

Ogunquit Lobster Pound, 504 Main Street, Ogunquit. 207-646-2516. ogunquitlobsterpound.com. This landmark serves lobster nine different ways, but most customers go the traditional route, choosing their lobster from an outdoor tank. The crustacean is then boiled in seawater and brought to their table.

Bob's Clam Hut, 315 Route 1, Kittery. 207-439-4233. bobsclamhut.com. Founded by Bob Craft in 1956, this small eatery has been cooking clams for generations of southern Mainers. The tasty little bivalves are served every which way — fried clams, clam strips, steamed clams, clam chowder, clam fritters, even clamburgers.

SHOP

L.L. Bean, 95 Main Street, Freeport. 877-755-2326. llbean.com. The famous outfitter sells more than the Maine hunting shoe invented by its founder Leon Leonwood Bean in 1912. At this four-store campus, shoppers find camping, hunting, and recreational equipment, clothing, and home furnishings. L.L. Bean has been a magnet for more than seventy factory outlet stores in Freeport.

DeLorme Map Store, Route 1, Freeport. 207-846-7100. delorme.com. DeLorme sells its wares in this shop dominated by Eartha, the world's largest rotating and revolving globe. You'll find atlases and gazetteers for every state, mapping software, GPS products, globes, and a large selection of travel guides.

Stonewall Kitchen, Route 1, York. 207-351-2712. stonewallkitchen.com. Stonewall Kitchen's flagship store offers samples of its specialty food items like jams, mustards, chutneys, vinegars, and barbecue sauces. The location has a café and cooking school.

All street names refer to Route 1 unless otherwise noted.